2013年

全国水利发展统计公报

2013 Statistic Bulletin on China Water Activities

中华人民共和国水利部 编

Ministry of Water Resources, People's Republic of China

图书在版编目（CIP）数据

2013年全国水利发展统计公报 = 2013 statistic bulletin on China water activities / 中华人民共和国水利部编. -- 北京：中国水利水电出版社，2014.11
 ISBN 978-7-5170-2647-1

Ⅰ. ①2… Ⅱ. ①中… Ⅲ. ①水利建设－经济发展－中国－2013 Ⅳ. ①F426.9

中国版本图书馆CIP数据核字(2014)第249321号

书　名	2013年全国水利发展统计公报 2013 Statistic Bulletin on China Water Activities
作　者	中华人民共和国水利部　编 Ministry of Water Resources, People's Republic of China
出版发行	中国水利水电出版社 （北京市海淀区玉渊潭南路1号D座　100038） 网址：www.waterpub.com.cn E-mail：sales@waterpub.com.cn 电话：（010）68367658（发行部）
经　售	北京科水图书销售中心（零售） 电话：（010）88383994、63202643、68545874 全国各地新华书店和相关出版物销售网点
排　版	中国水利水电出版社微机排版中心
印　刷	北京瑞斯通印务发展有限公司
规　格	210mm×297mm　16开本　3.5印张　48千字
版　次	2014年11月第1版　2014年11月第1次印刷
印　数	0001—1000册
定　价	**28.00**元

凡购买我社图书，如有缺页、倒页、脱页的，本社发行部负责调换

版权所有·侵权必究

目 录

1 水利固定资产投资 …………………………………………… 1
2 重点水利建设 ………………………………………………… 5
3 主要水利工程设施 …………………………………………… 8
4 水资源利用与保护 …………………………………………… 13
5 防洪抗旱 ……………………………………………………… 15
6 水利改革与管理 ……………………………………………… 17
7 水利行业状况 ………………………………………………… 23

Contents

I. Investment in Fixed Assets 27

II. Key Water Projects Construction 32

III. Key Water Structures 35

IV. Water Resources Utilization and Protection 39

V. Flood Control and Drought Relief 40

VI. Water Management and Reform 42

VII. Current Status of Water Sector 49

  2013年是贯彻落实党的十八大精神、实现良好开局的一年，是改革开放进程中具有里程碑意义的一年，也是水利建设继续快速推进、水利改革发展取得突出成效的一年。在党中央、国务院的正确领导下，广大水利干部职工迎难而上、真抓实干，推动各项水利工作取得新进展，为实现粮食产量"十连增"，进一步改善民生、促进经济持续健康发展和社会和谐稳定提供了有力的水利支撑与保障。

1 水利固定资产投资

2013年，全社会共落实水利建设投资计划3954.0亿元，较上年减少4.0%。其中，中央政府投资1762.6亿元，较上年减少17.4%；地方政府投资1726.9亿元，较上年增加6.1%；利用外资10.8亿元，较上年增加208.6%；国内贷款194.2亿元，较上年增加3.8%；企业和私人投资147.8亿元，较上年增加42.7%；其他投资111.7亿元，较上年增加82.5%。

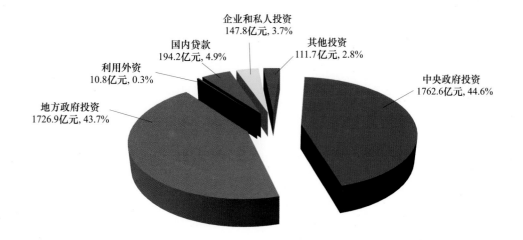

全社会水利固定资产投资计划

在中央水利建设投资中，水利部门投资1408.3亿元，南水北调水利工程建设投资281.8亿元，其他部门水利建设投资72.5亿元。水利部门投资按投资来源分，中央预算内固定资产投资717.1亿元、水利建设基金22.3亿元、财政专项资金669.0亿元；按资金投向分：防洪工程投资619.0亿元，占43.95%；水资源工程投资688.2亿元，占48.87%；水土保持及生态工程投资78.2亿元，占5.55%；专项工程投资23.0亿元，占1.63%。

当年正式施工的水利建设项目20266个，在建项目投资总规模15346.0亿元，较上年增加12.0%。其中，有中央投资的水利建设项目10974个，较上年增加40.0%；在建投资规模9764.0亿元，较上年增加21.8%。新开工项目12199个，较上年减少8.7%，新增投资规模4645.9亿元，比上年增加38.8%。

全年水利建设完成投资3757.6亿元，较上年减少206.6亿元，减少5.2%。其中：建筑工程完成投资2782.8亿元，较上年增加1.7%；安装工程完成投资173.6亿元，较上年减少27.0%；机电设备及工器具购置完成投资161.1亿元，较上年减少9.5%；其他完成投资（包括移民征地补偿等）640.2亿元，较上年减少21.1%。

	2007年/亿元	2008年/亿元	2009年/亿元	2010年/亿元	2011年/亿元	2012年/亿元	2013年/亿元	增加比例/%
全年完成	944.9	1088.2	1894.0	2319.9	3086.0	3964.2	3757.6	-5.2
建筑工程	672.5	781.5	1297.2	1524.9	2103.2	2736.5	2782.8	1.7
安装工程	46.5	67.4	113.4	109.6	121.7	237.8	173.6	-27.0
设备及各类工器具购置	56.8	60.0	125.0	124.5	115.2	178.1	161.1	-9.5
其他（包括移民征地补偿等）	169.1	179.3	358.4	560.9	745.9	811.8	640.2	-21.1

在全年完成投资中，防洪工程建设完成投资1335.8亿元，水资源工程建设完成投资1733.1亿元，水土保持及生态工程完成投资102.9亿元，水电、机构能力建设等专项工程完成投资585.8亿元。七大江河流域完成投资3284.4亿元，东南诸河、西北诸河以及西南诸河等其他流域完成投资473.2亿元。东部、东北、中部、西部地区完成投资分别为1117.2亿元、211.2亿元、1056.0亿元、1373.2亿元，占全部完成投资的比例分别为29.7%、5.6%、28.1%和36.5%。

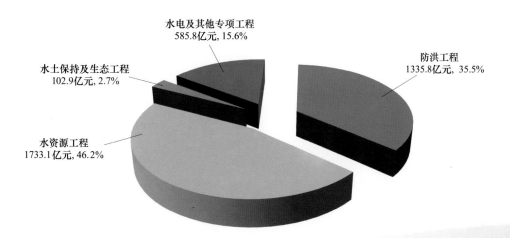

2013年分用途完成投资情况

在全年完成投资中，中央项目完成投资430.4亿元，地方项目完成投资3327.2亿元。大中型项目完成投资905.9亿元，小型及其他项目完成投资2851.7亿元。各类新建工程完成投资2555.3亿元，扩建、改建等项目完成投资1202.3亿元。

全年水利建设新增固定资产2780.4亿元。截至2013年年底，在建水利项目累计完成投资10142.3亿元，投资完成率为61.3%，比上年提高2.8个百分点；累计新增固定资产5577.1亿元，固定资产形成率为55.0%，比上年减少9.8个百分点。

全年水利建设完成土方、石方和混凝土方分别为36.0亿立方米、5.4亿立方米、0.7亿立方米。截至2013年年底，在建项目计划实物工程量完成率分别为：土方71.2%、石方56.7%、混凝土方66.9%。

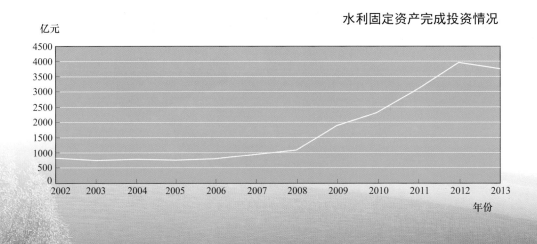

水利固定资产完成投资情况

2 重点水利建设

江河湖泊治理。全年在建江河湖泊治理工程 3977 处,其中,堤防建设 427 处,大江大河及重要支流治理 498 处,中小河流治理 2921 处,行蓄洪区安全建设及其他项目 131 处。截至 2013 年年底,在建项目累计完成投资 1791.1 亿元,投资完成率 59.2%。全年实施河道整治长度 11009.3 公里,当年完成 8969.7 公里。国家正式启动淮河进一步治理工作,累计安排投资 131.2 亿元,截至 2013 年年底完成 93.6 亿元,投资计划完成率 70.9%。

水库及枢纽工程建设。全年在建枢纽工程 348 座,截至 2013 年年底,在建项目累计完成投资 1553.9 亿元,项目投资完成率 65.7%。全年在建病险水库除险加固工程 4530 座,截至 2013 年年底,在建项目累计完成投资 350.9 亿元,项目投资完成率 82.0%。当年在建重点小型病险水库除险加固工程 4317 处,截至 2013 年年底,基本完成除险加固任务的小型病险水库 15887 座。

水资源配置工程建设。全年水资源配置工程建设在建投资规模3304.4亿元，累计完成投资2503.9亿元，项目投资完成率75.8%。南水北调东、中线一期工程在建规模2493.4亿元，累计完成投资2434.1亿元，全年完成投资404.9亿元。加快建设河北双峰寺、吉林引松供水、浙江舟山大陆引水二期、云南牛栏江-滇池补水等23处大中型引调水工程。

农村水利建设。全年农村饮水安全工程在建投资规模796.6亿元，累计完成投资691.2亿元。全年解决6343万农村居民和农村学校师生的饮水安全问题。截至2013年年底，农村集中式供水受益人口比例达到73.1%。当年中央安排预算内投资107.3亿元，用于规划内211处大型灌区续建配套与节水改造、19处三江平原新建灌区建设、153处中型灌区节水配套改造建设、14个省份大型灌排泵站更新改造、规模化节水灌溉增效示范和牧区水利项目建设，安排中央财政资金243.3亿元用于小型农田水利设施建设。全年新增有效灌溉面积1552千公顷，新增节水灌溉面积2406千公顷。

农村水电建设。全年全国农村水电站建设完成投资198亿元，新增水电站389座，装机容量246万千瓦。全国农村水电配套电网建设共完成投资65亿元，新增110千伏及以上变电站容量472万千伏安；新增35（63）千伏变电站容量129万千伏安；配电变压器容量443万千伏安。新投产10千伏及以上高压线路2.5万公里，低压线路4.3万公里。

水土流失治理。全年水土保持及生态工程在建规模306.3亿元，累计完成投资227.4亿元。全国新增水土流失综合治理面积5.3万平方公里，其中国家水土保持重点工程新增水土流失治理面积1.32万平方公里。全年新增封育保护面积1.68万平方公里。实施2153条小流域水土流失综合治理，兴建黄土高原淤地坝285座，治理崩岗2000多处。全年新修基本农田727千公顷（其中：新增梯田553千公顷，坝地16千公顷），新栽种水保林面积1411千公顷，新增种草面积340千公顷。开展国家重点治理的项目县700多个。坡耕地水土流失综合治理试点工程建设范围扩大到22个省、160个县，全年完成坡改梯72.67千公顷。

行业能力建设。全年水利行业能力建设完成投资61.1亿元。其中：防汛通信设施投资19.5亿元，水文建设投资20.6亿元，科研教育设施投资1.5亿元，水利前期投资16.7亿元，其他投资2.8亿元。

3 主要水利工程设施

堤防和水闸。 全国已建成五级以上江河堤防27.68万公里❶，累计达标堤防17.98万公里，堤防达标率为65.0%；其中一级、二级达标堤防长度为2.95万公里，达标率为76.8%。全国已建成江河堤防保护人口5.7亿人，保护耕地4.3万千公顷。全国已建成流量为5立方米每秒及以上的水闸98192座，其中大型水闸870座；其中：分洪闸7985座，排（退）水闸17509座，挡潮闸5834座，引水闸11106座，节制闸55758座。

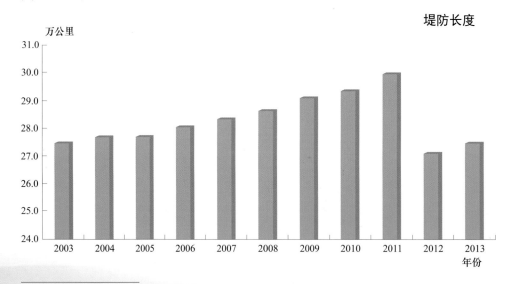

---

❶ 2011年以前各年堤防长度含部分地区五级以下江河堤防。

水库和枢纽。全国已建成各类水库97721座,水库总库容8298亿立方米。其中:大型水库687座,总库容6528亿立方米,占全部总库容的78.7%;中型水库3774座,总库容1070亿立方米,占全部总库容的12.9%。全国大中型水库大坝安全达标率为96.8%。

农业灌溉。全国设计灌溉面积大于2000亩及以上的灌区共22387处,耕地灌溉面积33928千公顷。其中:50万亩以上灌区176处,耕地灌溉面积6241千公顷;30万~50万亩大型灌区280处,耕地灌溉面积5010千公顷。截至2013年年底,全国耕地灌溉面积63473千公顷,占全国耕地面积的52.9%。全国节水灌溉工程面积27109千公顷,其中:喷、微灌面积6847千公顷,低压管灌面积7424千公顷。

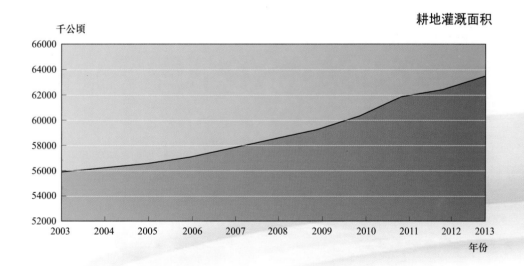

机电井和泵站。截至2013年年底,全国已累计建成日取水大于等于20立方米的供水机电井或内径大于200毫米的灌溉机电井共458.4万眼。全国已建成各类装机流量1立方米每秒或装机功率50千瓦以上的泵站90650处,其中:大型泵站366处,中型泵站4161处,小型泵站86123处。

农村水电。截至2013年年底,全国共建成农村水电站46849座,装机容量7119万千瓦,占全国水电装机容量的25.4%。全国农村水电年发电量2233亿千瓦时,占全国水电发电量的24%。全年累计解决47万无电人口用电问题。

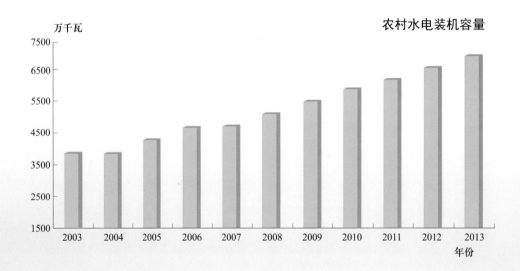

水土保持工程。截至 2013 年年底，全国水土流失综合治理面积达 106.89 万平方公里，累计封禁治理保有面积达 77 万平方公里，建成生态清洁型小流域 160 条。全国水土保持监测网络和信息系统建设二期工程竣工验收，建成了淮河、松辽、珠江、太湖 4 个流域机构监测中心站，建成 18 个省级监测总站、75 个监测分站、715 个监测点，建设开发了水土保持数据库和应用系统等。

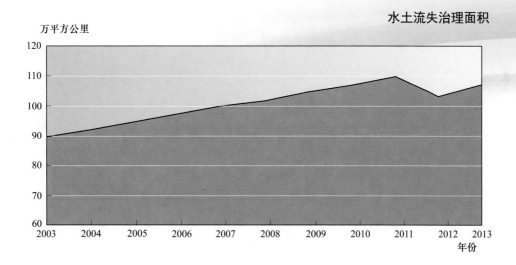

水土流失治理面积

水文和信息化。截至 2013 年年底，全国共有各类水文测站 86554 处，其中：国家基本水文站 3195 处，专用水文站 816 处，水位站 9330 处，雨量站 43028 处，水质站 11795 处，地下水监测站 16407 处，蒸发站 14 处，实验站 57 处，墒情站 1912 处。全国共有拍报水情的水文测站 24518 处，发布预报的水文站 1207 处。已建成水环境监测（分）中心 286 个，水质监测基本覆盖了全国主要江河湖库。

水利信息化建设进入全方位、多层次推进的新阶段。省级以上水利部门接入水利信息网络的各种类型 PC 机数量达到 74964 台，服务器设备 3273 套；省级以上水利部门已配备的各类在线存储设备的存储能力 2057971.9GB。省级以上水利部门可接受信息的各类水利信息采集点 105930 个，其中自动采集点 63461 个，正常运行的数据库达 706 个，存储的数据量达到 333528.7GB。

4 水资源利用与保护

2013年全国水资源总量27957.9亿立方米，比常年值偏多0.9%；全国平均降水量661.9毫米，比常年值偏多3.0%，较上年减少3.8%。截至2013年年底，全国监测的588座大型水库年末蓄水总量3005.4亿立方米，比2013年年初减少232.9亿立方米；3271座中型水库年末蓄水量为395.3亿立方米，比2013年年初减少7.1亿立方米。

全年全国总供水量6183.4亿立方米，其中地表水源占81.0%，地下水源占18.2%，其他水源占0.8%。全国总用水量6183.4亿立方米，其中：生活用水750.1亿立方米（其中城镇生活用水占75.5%），占总用水量的12.1%；工业用水1406.4亿立方米，占总用水量的22.8%；农业用水3921.5亿立方米，占总用水量的63.4%；生态环境补水105.4亿立方米，占总用水量的1.7%。与2012年比较，用水量增加52.2亿立方米，其中：生活用水增加10.4亿立方米，工业用水增加25.7亿立方米，农业用水增加19.0亿立方米，生态环境补水减少2.9

亿立方米。全国人均用水量为456立方米。万元GDP用水量109立方米（当年价），比2012年减少7.6%；万元工业增加值用水量67立方米（当年价），比2012年减少5.3%。

根据对全国20.8万公里河流水质评价结果，水质符合和优于Ⅲ类水的河长占总评价河长的68.6%。与2012年相比，水质为Ⅰ~Ⅲ类水河长比例上升0.6个百分点，水质为劣Ⅴ类水河长比例下降0.8个百分点。

5 防洪抗旱

2013年，全国洪涝灾害总体偏轻。全国农作物受灾面积11901千公顷，成灾面积6623千公顷，受灾人口1.2亿人，因灾死亡775人、失踪374人，倒塌房屋53万间，县级以上城市受淹243座，直接经济损失约3146亿元，其中水利设施直接经济损失445亿元。东北地区、四川、广东等地受灾较重。全国因山洪灾害造成人员死亡占全部死亡人数的72%，因台风造成经济损失约占全国洪涝灾害直接经济损失的40%，为2006年以来最重。

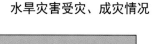

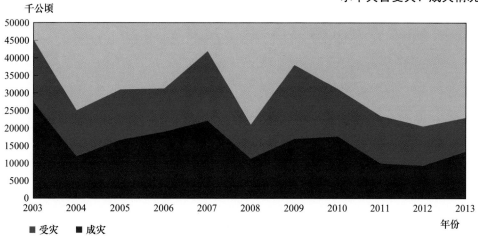

水旱灾害受灾、成灾情况

全国旱灾总体偏轻。江淮、江南、江汉部分地区主汛期高温少雨伏旱严重。全国农田因旱受灾面积11220千公顷，成灾面积6971千公顷，直接经济总损失1275亿元。全国因旱累计有2241万城乡人口、1179万头大牲畜发生临时性饮水困难。

全年中央下拨用于防汛抗旱方面的经费共98.89亿元，其中：水利建设基金5.92亿元，特大防汛经费25.87亿元，特大抗旱经费16.26亿元，山洪灾害防治经费43亿元，抗旱服务队补助7.84亿元。

2013年，防洪减淹耕地3967千公顷，避免城市进水受淹156座，防洪减灾经济效益2358亿元。解决了2007万城乡居民和936万头大牲畜因旱临时饮水困难，全年完成抗旱浇地面积24667千公顷，抗旱挽回粮食损失399亿公斤。

6 水利改革与管理

水利规划和前期工作。2013年,中央层面批复各类水利规划26项,其中:国务院批复7项、国家发展和改革委员会批复6项、水利部批复13项。七大流域综合规划修编全部得到国务院批复,其中2013年国务院批复了黄河、淮河、海河、珠江、松花江、太湖流域综合规划;全国水利发展"十三五"规划编制准备工作正式启动,全国现代灌溉发展规划、全国水中长期供求规划、全国水资源保护规划、《全国抗旱规划实施方案(2014—2016)》、全国江河重要河道采砂管理规划、全国地下水利用与保护规划、全国水土保持规划、全国河口海岸滩涂开发管理规划等一批重要规划进展顺利;全国牧区水利发展规划编制完成,全国水利现代化规划纲要、全国治涝规划等编制工作相继启动。主要江河流域(第一批25条河)水量分配方案已获得初步成果。全年共向国家发展和改革委员会报送46项重点水利项目,投资规模1461.49亿元,国家发展和改革委员会批复重点水利项目39项,总投资1077.33亿元。其中:项目建议书12项,总投资428.06亿元;可行性研究18项,总投资451.12亿元;初步设计14项,总投资288.31亿元。

水利立法与水政管理。2013年水利部共计准予（延续）水行政许可1420件，其中：建设项目水资源论证机构资质认定119件，建设项目水资源论证报告审批2件，水利工程建设监理单位资质认定355件，水利工程质量检测单位资质认定159件，生产建设项目水土保持监测单位资质认定42件，水利水电建设项目环境影响报告书（表）预审19件，生产建设项目水土保持方案审批285件，生产建设项目水土保持设施验收179件，水文、水资源调查评价机构资质认定172件，水利工程启闭机使用许可证核发72件。长江中下游干流河道采砂规划划定41个可采区，2013年度审批采砂经验许可权32个，许可年度采砂总量9605.8万吨，许可采砂船只数量190艘。全国共查处水事违法案件55437件，已结案50036件，结案率90.26%，挽回直接经济损失18299万元。全国共调处水事纠纷5537件，挽回经济损失4815万元。2013年办结行政复议案件52件。

水务管理。全国组建水务局或由水利局承担水务管理职能的县级以上行政区共计1942个，占全国县级以上行政区总数的79.7%。在组建的1479个水务局中，省级水务局（厅）4个，副省级水务局7个，地级水务局213个，县级水务局1255个。水务系统共有自来水厂4039座，供水管道总长53.4万公里，自来水供水能力25452万立方米每日，年供水总量379.6亿立方米。污水处理厂2159座，排水管道总长26.3万公里，污水处理能力9365万立方米每日，年污水处理总量254.1亿立方米。水务系统共有水务企业2761家，年末固定资产总值2389.6亿元，年销售收入504.4亿元，年利润4.3亿元。水务系统城市水务投资总额1224.4亿元。全国城市（县城）水源地合计4566个，水源年供水

能力1124.0亿立方米。全国城市（县城）年污水处理回用量48.2亿立方米，除污水处理回用外的其他非传统水资源利用量694.7亿立方米。

建设与管理改革。全国水利工程管理体制改革基本完成并通过验收。各地纳入水管体制改革范围的水管单位14325个，经精简撤并调整为12979个，较改革前下降9%。12979个水管单位共落实两项经费129.6亿元，落实率83%，其中：落实公益性人员基本支出103.9亿元，落实率92%；落实公益性工程维修养护经费75.7亿元，落实率73%。实行管养分离（包括内部管养分离）的水管单位8995个，占水管单位总数的68.5%。全年新批准取得水利工程施工监理专业甲级资质单位22个，乙级资质单位59个，丙级资质单位187个；取得水土保持工程施工监理专业甲级资质单位9个，乙级资质单位8个，丙级资质单位20个；取得机电及金属结构设备制造监理专业乙级资质单位2个；取得水利工程建设环境保护监理专业资质（不分级）单位8个。全年新增取得岩土工程类质量检测甲级资质单位17个，取得混凝土工程类质量检测甲级资质单位15个，取得金属结构类质量检测甲级资质单位3个，取得机械电气类质量检测甲级资质单位1个，取得量测类质量检测甲级资质单位5个。截至2013年年底，累计批准国家级水利风景区588个，其中：水库型296个，自然河湖型117个，城市河湖型94个，湿地型35个，灌区型24个，水土保持型22个。

农村水利改革。全国成立的以农民用水户协会为主要形式的农民用水合作组织累计达到8.05万多家，协会管理灌溉面积约2.62亿亩，占全国耕地灌溉面积的28%。全国已建立基层水利服务机构2.9万个，落实人员13万多人。继续推进以落实"两费"为主要内容的灌区管理体制改革。大型灌区管理单位公益性人员基本支出和公益性工程维修养护经费落实率分别达到60%和41%。

水土保持管理。全国共审批开发建设项目水土保持方案3.1万个，其中水利部审批国家大中型项目水土保持方案285个，涉及防治责任范围14894平方公里。全年完成生产建设项目的水土保持设施验收6611个。截至2013年年底，12个省（自治区、直辖市）出台了水土保持法省级实施办法或条例。

水价改革。2013年，全国农业供水成本25.68分每立方米，其中：国有水利工程农业供水成本17.26分每立方米，末级渠系供水成本8.42分每立方米。农业水价为9.14分每立方米，其中：国有水利工程农业水价6.89分每立方米，末级渠系农业水价2.25分每立方米，约为成本的35.59%，全国平均农业水费实收率为86.48%。据不完全统计，实行水务管理地区原水水价0.01~6.0元每立方米，地表水征收水资源费0.002~2.0元每立方米，地下水征收水资源费0.01~10.0元每立方米。

水电改革和管理。全国17个省（自治区、直辖市）开展了水能资源使用权有偿出让，16个省（自治区、直辖市）出台了水能资源管理的规范性文件。15 个省明确由水行政主管部门负责水能资源统一管理。积极落实农村水电安全生产"双主体"责任,安全监管覆盖率超过 99%。

水利安全监督。水利行业共发生 15 起生产安全事故,死亡 24 人,水利安全生产形势保持总体平稳态势。全年共派出 149 个稽查组,稽查项目 429 个,查出各类问题 2229 个,下发稽查整改意见 301 份,对 5 个省存在突出问题的 5 个项目进行了全国通报。组织流域机构派出 31 个稽查组,对 93 个存在突出问题的项目整改情况进行重点复查和督办,同时对 2011 年以来水利部稽查的 8 个省 117 个项目的整改情况进行全面核查和评估。指导和推动省级水行政主管部门开展水利稽查工作,各地完成稽查项目 1308 个,发现问题 9215 个,印发整改意见 960 份。

水利移民。据不完全统计,2013 年在建的大中型水库 114 座,涉及 19 个省(自治区、直辖市),搬迁人口 11.7 万人,开工建设安置点 367 个,安置农村移民 9.4 万人,新建住房 330 万平方米,调整土地 9.6 万亩。

水利科技。全年共安排 4.2 亿元资金用于水利科技项目,其中:组织立项国家科技支撑计划 3 项,水利公益性行业科研专项 66 项,"948"计划、国家农业科技成果转化资金专项、水利部科技成果重点推广计划等各类科技计划项目60项。水利科技项目成果获国家科技进步奖4项。截至2013年年底，水利系统共有国家级和部级重点实验室12个，工程技术研究中心13个。落实中央级科学事业单位修缮购置专项资金10844.1万元。水利行业现行有效标准达767项，在编（含修订）水利技术标准194项，列入《水利技术标准体系表》拟编水利技术标准81项。

国际合作。成功举办或参与多双边国际交流活动37次，签署双边水利合作协议1份。组织召开双边政府固定交流机制会议7次。世行贷款淮河流域重点平原洼地排涝治理项目进展顺利，贷款金额2亿美元。正在执行的政府间合作项目5个，获得立项的国家国际科技合作专项项目3个，专项资金897万元；正在执行的国家国际科技合作专项项目7个，专项资金1927万元。

7 水利行业状况

职工与工资。全国水利系统从业人员104万人,比上年减少2.9%。其中,全国水利系统在岗职工100.5万人,比上年减少2.8%。在岗职工中,部直属单位在岗职工7.0万人,比上年减少5.4%,地方水利系统在岗职工93.5万人,比上年减少2.6%。全国水利系统在岗职工工资总额为415.3亿元,比上年增加6.7%。全国水利系统在岗职工年平均工资41453元,比上年增加10.0%。

职工与工资情况

	2003年	2004年	2005年	2006年	2007年	2008年	2009年	2010年	2011年	2012年	2013年
在岗职工人数/万人	122.9	118.2	110.5	109.2	106.8	105.6	103.7	106.6	102.5	103.4	100.5
其中:部直属单位/万人	6.4	6.4	6.6	6.8	7.2	7.2	7.2	7.4	7.5	7.4	7.0
地方水利系统/万人	116.5	111.8	103.9	102.3	99.6	98.4	96.5	96.3	95.0	96.0	93.5
在岗职工工资/亿元	140.6	157.1	159.8	184.3	211.28	234.37	264.74	297.91	351.37	389.1	415.3
年平均工资/(元/人)	11443	13054	13969	16776	19573	22143	25633	28816	34283	37692	41453

勘察设计。2013年全国具有水利行业设计甲级资质的单位108家，拥有设计乙级资质的单位361家，丙级资质1141家，职工总人数近8万人。

水利建设。全国共有水利水电工程施工总承包特级资质企业10家，水利水电工程施工总承包一级资质企业204家。注册水利工程建设监理工程师38767人，一级注册建造师水利水电工程专业资格13525人。

全国水利发展主要指标（2008—2013年）

指标名称	单位	2008年	2009年	2010年	2011年	2012年	2013年
1. 灌溉面积	千公顷	64120	65165	66352	67743	67780	69481
2. 耕地灌溉面积	千公顷	58472	59261	60348	61682	62491	63473
其中：本年新增	千公顷	1318	1533	1722	2130	2151	1552
3. 节水灌溉面积	千公顷	24436	25755	27314	29179	31217	27109
4. 万亩以上灌区	处	5851	5844	5795	5824	7756	7709
其中：30万亩以上	处	325	335	349	348	456	456
万亩以上灌区耕地灌溉面积	千公顷	29440	29562	29415	29748	30087	30216
其中：30万亩以上	千公顷	15401	15575	15658	15786	11260	11252
5. 当年解决农村饮水安全人口	万人	5378	7295	6717	6398	7294	6343
6. 除涝面积	千公顷	21425	21584	21692	21722	21857	21943
7. 水土流失治理面积	万平方公里	101.6	104.3	106.8	109.7	103.0	106.9
8. 水库	座	86353	87151	87873	88605	97543	97721
其中：大型水库	座	529	544	552	567	683	687
中型水库	座	3181	3259	3269	3346	3758	3774
水库总库容	亿立方米	6924	7064	7162	7201	8255	8298

续表

指标名称	单位	2008年	2009年	2010年	2011年	2012年	2013年
其中：大型水库	亿立方米	5386	5506	5594	5602	6493	6528
中型水库	亿立方米	910	921	930	954	1064	1070
9. 全年水利工程总供水量	亿立方米	5910	5933	6022	6107	6142	6183
10. 堤防长度	万公里	28.7	29.1	29.4	30.0	27.17	27.68
保护耕地	千公顷	45712	46547	46831	45418	42597	42573
堤防保护人口	万人	57289	58978	59853	59697	56566	57138
11. 水闸总计	座	41626	42523	43300	44306	97256	98192
其中：大型水闸	座	504	565	567	599	862	870
12. 年末全国水电装机容量	万千瓦	17090	19686	21157	23007	24881	28026.0
全年发电量	亿千瓦时	5614	5055	6813	6507	8657	9304.2
13. 农村水电装机容量	万千瓦	5127.4	5512.1	5924.0	6212.3	6568.6	7118.6
全年发电量	亿千瓦时	1628	1567	2044	1757	2173	2233
14. 当年完成水利建设投资	亿元	1088.2	1894.0	2319.9	3086.0	3964.2	3757.6
按投资来源分：							
（1）预算内拨款	亿元	390.4	929.9	918.1	898.8	1291.5	1136.8
（2）预算内专项	亿元	160.5	128.8	94.8	29.2	25.4	12.7
（3）财政专项	亿元	—	—	—	564.8	1004.2	937.5
（4）水利建设基金	亿元	60.5	105.5	215.2	79.6	120.0	108.1
（5）重大水利工程建设基金	亿元	—	—	—	437.8	434.9	425.0
（6）土地出让收益	亿元	—	—	—	12.1	25.8	30.6
（7）水资源费	亿元	—	—	—	18.2	21.5	38.5
（8）国内贷款	亿元	96.9	152.9	337.4	270.3	265.6	172.7
（9）利用外资	亿元	10.5	7.6	1.3	4.4	4.1	8.6
（10）自筹资金	亿元	235.4	333.9	316.2	406.8	350.4	360.7
（11）企业和私人投资	亿元	35.9	41.4	48.0	74.9	113.4	160.7
（12）债券	亿元	—	—	2.5	3.9	5.2	1.7
（13）其他投资	亿元	98.1	194.2	386.5	285.1	302.4	364.0

续表

指标名称	单位	2008年	2009年	2010年	2011年	2012年	2013年
按投资用途分：							
（1）防洪工程	亿元	370.1	674.8	684.6	1018.3	1426.0	1335.8
（2）水资源工程	亿元	467.8	866	1070.5	1284.1	1911.6	1733.1
（3）水土保持及生态建设	亿元	76.9	86.7	85.9	95.4	118.1	102.9
（4）水电工程	亿元	77.4	72	105.4	109.0	117.2	164.4
（5）行业能力建设	亿元	10.6	10.6	19.6	40.2	59.6	52.5
（6）前期工作	亿元	16.0	15.9	24.9	42.0	40.7	40.7
（7）其他	亿元	69.4	167.9	329.1	496.9	291.1	328.2

说明：1. 本公报不包括香港特别行政区、澳门特别行政区以及台湾省的数据。
2. 节水灌溉面积2013年统计数据与第一次全国水利普查数据进行了衔接，其他水利发展主要指标2012年统计数据已与第一次全国水利普查数据进行了衔接。其中，堤防长度与水利普查成果衔接后，进一步明确为5级及以上堤防。
3. 2011年及以前万亩以上灌区处数及灌溉面积按有效灌溉面积达到万亩进行统计，2012年以来按设计灌溉面积达到万亩以上进行统计。
4. 农村水电的统计口径为装机容量5万千瓦及5万千瓦以下水电。

2013 STATISTIC BULLETIN ON CHINA WATER ACTIVITIES

Ministry of Water Resources, P. R. China

In the year of 2013, we got off to a good start of implementing the essential ideas of the 18th national congress proposed by the Communist Party of China (CPC). This year is of milestone significance to the progress of reform and opening-up of China and had made outstanding achievements in continuously sustaining large scale of investment for water project construction as well as reform and development of the water sector. Under the wise leadership of the Central Committee of CPC and the State Council, the water departments at all levels have braved difficulties and conducted solid work for new progress of water development in all aspects. The support provided by the water sector has ensured the realization of "ten consecutive growth" of grain production, enabled further enhancement of people's well being, and kept healthy and continuous development of economy as well as social stability and harmony in China.

I. Investment in Fixed Assets

In 2013, the total investment for water project construction from the whole society was 395.40 billion Yuan, a 4.0% decrease comparing to the year of 2012. Divided by sources, 176.26 billion Yuan was financed by the central government with a decrease of 17.4%, 172.69 billion Yuan financed by local governments, 6.1% increase; 1.08 billion Yuan of foreign investment, 208.6% increase; 19.42 billion Yuan of domestic loans, 3.8% increase; 14.78 billion Yuan from enterprises and private sector, 42.7% increase; and 11.17 billion Yuan from other sources, 82.5% increase.

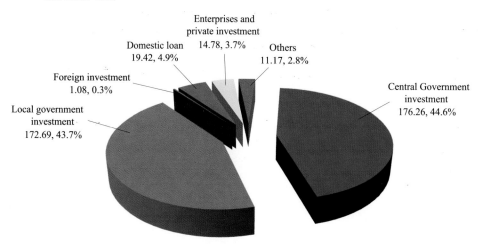

Total fixed assets investment plan of water sector

The total investment of Central Government for water project construction was 176.26 billion Yuan in 2013, among which 140.83 billion Yuan from MWR, 28.18 billion Yuan from construction funds of South – North Water Diversion Project and 7.25 billion Yuan from investments of other ministries; The total investment of MWR was 140.83 billion Yuan, including 71.71 billion Yuan of fixed assets investment of Central Government Budget, 2.23 billion Yuan of Water Construction Funds and 66.90 billion Yuan of Special Funding of Central Government Budget; Divided by the types of projects: 61.90 billion Yuan for flood control, taking 43.95%; 68.82 billion Yuan for water resources project, taking 48.87%; 7.82 billion Yuan for soil and water conservation and ecological improvement, taking 5.55%; and 2.30 billion Yuan for special projects, taking 1.63%.

A total of 20,266 water projects were under construction in 2013, with a total investment of 1534.60 billion Yuan, with an increase of 12.0% comparing to that of the year before. The projects with Central Government finance were 10,974 with an increase of 40.0% comparing to the year before. The total funds used by projects under construction reached 976.40 billion Yuan and increased 21.8% comparing to the year before. There were 12,199 newly-constructed projects in 2013, with a decrease of 8.7% and newly-added investment was 464.59 billion Yuan with a increase of 38.8%.

Completed investment for water project construction in 2013 amounted to 375.76 billion Yuan, with a decrease of 20.66 billion Yuan or 5.2% decrease comparing to the year before. In which, 278.28 billion Yuan put into construction project with a 1.7% increase; 17.36 billion Yuan for installation with a decrease of 27.0%; 16.11 billion Yuan for purchase of machinery, electric equipment and instruments, with an decrease of 9.5%; and 64.02 billion Yuan for other purposes (including compensation of resettlement and land acquisition), with a decrease of 21.1%.

	2007 /billion Yuan	2008 /billion Yuan	2009 /billion Yuan	2010 /billion Yuan	2011 /billion Yuan	2012 /billion Yuan	2013 /billion Yuan	increase /%
Yearly Completed	94.49	108.82	189.40	231.99	308.60	396.42	375.76	−5.2
Construction project	67.25	78.15	129.72	152.49	210.32	273.65	278.28	1.7
Installation project	4.65	6.74	11.34	10.96	12.17	23.78	17.36	−27.0
Procurement of instruments and equipment	5.68	6.00	12.50	12.45	11.52	17.81	16.11	−9.5
Others (including compensation for resettlement and land expropriation)	16.91	17.93	35.84	56.09	74.59	81.18	64.02	−21.1

In the total completed investment, 133.58 billion Yuan was allocated to the construction of flood control projects, 173.31 billion Yuan for the construction of water resources projects, 10.29 billion Yuan for soil and water conservation and ecological restoration, and 58.58 billion Yuan for special projects, such as hydropower development and capacity building.

The completed investment for seven major river basins reached 328.44 billion Yuan, of which 47.32 billion Yuan was invested in river basins in the southeast, southwest and northwest of China. Moreover, completed investments in eastern, northeast, middle and western regions were 111.72 billion Yuan, 21.12 billion Yuan, 105.60 billion Yuan and 137.32 billion Yuan respectively, accounting 29.7%, 5.6%, 28.1%, and 36.5% of the total.

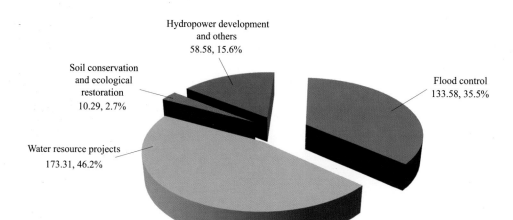

Completed investment of projects in 2013

Of this total competed investment, the Central Government contributed 43.04 billion Yuan, and local governments contributed 332.72 billion Yuan. Large and medium sized projects completed investment of 90.59 billion Yuan; small-sized and other projects completed 285.17 billion Yuan. Newly-constructed project completed 255.53 billion Yuan; and reconstruction and expansion completed 120.23 billion Yuan.

In 2013, the newly-added fixed assets totaled 278.04 billion Yuan. By the end of 2013, the accumulated investment in projects under construction was 1,014.23 billion Yuan, and the rate of completed investment reached 61.3%, an increase of 2.8% Comparingto 2012. The newly increased fixed assets of projects under-construction valued 557.71 billion Yuan, and the rate of investment transferred into fixed assets was 55.0%, an decrease of 9.8% comparing to 2012.

In 2013, the completed civil works of earth, stone and concrete structures were 3.60 billion m^3, 540 million m^3, and 70 million m^3 respectively. By the end of 2013, the ratio of complete quantity of earthwork, stonework, concrete of the under-construction projects were 71.2%, 56.7%, and 66.9% respectively.

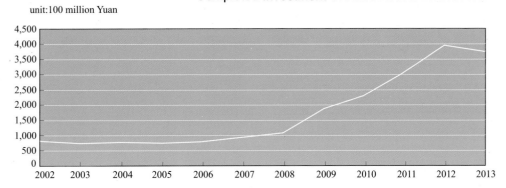

Completed Investment of Fixed Assets Investment

II. Key Water Projects Construction

Harness of rivers and lakes. In 2013, there were 3,977 river harness projects under construction that had spent 179.11 billion Yuan, accounting for 59.2% of the total completed investment. There were 11,009 km of river channels trained and 8,969.7 km of which completed. Among 3,977 projects, there was 427 embankments projects, 498 large rivers projects, 2,921 middle-small river harness projects, others was 131. Implementation Plan for Furthering Huaihe River Harness was jointly issued by the National Development and Reform Commission and Ministry of Water Resources in 2013. A total of 13.12 billion Yuan was allocated according to the investment plan and 9.36 billion Yuan was actually completed with a completion rate of 70.9%.

Reservoir projects. There were 348 reservoir projects under construction in 2013, with an accumulative investment of 155.39 billion Yuan, accounting for 65.7% of the total completed investment. There were 4,530 hazard vulnerable reservoirs completing repair or reinforcement, with an accumulated investment of 35.09 billion Yuan, accounting for 82.0% of the total completed investment. The Central Government spending for reinforcement of large and medium reservoirs as well as some small reservoirs of great significance reached to 4,317 projects. The plan for reinforcement of major small reservoirs that are vulnerable to hazard was initiated and 15,887 reservoirs had completed reinforcement by the end of 2013.

Water allocation projects. The yearly investment for water allocation projects reached to 330.44 billion Yuan. The completed investment in these projects had accumulated to 250.39 billion Yuan, accounting for 75.8% of the total completed investment. The scale of under-constructed projects of phase-I of eastern and middle routes of South-North Water Diversion Project reached to 249.34 billion Yuan. The accumulated completed investment was 243.41 billion Yuan and the completed investment in 2013 was 40.49 billion Yuan. Water source projects, including Shuangfengsi in Hebei Province, Water Diversion from Songhua River in Jilin Province, Water Diversion II from Zhoushan in Zhejiang Province and Water Diversion from Niulan River to Dianchi Lake in Yunnan Province, all 23 projects accelerated pace of construction.

Irrigation, drainage and rural water supply. The investment to the under-constructed projects for providing safe drinking water reached to 79.66 billion Yuan, with an accumulated investment of 69.12 billion Yuan. The newly increased capacity helped 63.43 million people access to safe drinking water. By the end of 2013, the beneficial rural population who have centralized water supply made up a percentage of 73.1% of the total. The Central Government allocated 10.73 billion Yuan for completion of counterpart systems and rehabilitation of 211 large irrigation districts for water conservation and water saving, newly-construction of 19 irrigation districts in Sanjiang Plain, rehabilitation and continuous construction of counterpart systems for 153 medium irrigation districts for water saving purpose, rehabilitation of large irrigation and drainage pumping stations in 14 provinces, construction of demonstration projects for large-scale extension and benefit increase of water-saving irrigation systems as well as pilot projects in pastureland. In addition, 24.33 billion Yuan from central government finance invested to build small-scale farmland waterworks for irrigation and drainage and water supply in rural areas. The newly-added effective irrigated area reached 1,552,000 ha; moreover, new-added water-saving irrigated area was 2,406,000 ha.

Rural hydropower and electrification. In 2013, the completed investment of rural hydropower station construction amounted to 19.8 billion Yuan; the newly increased hydropower stations were 389, with a total installed capacity of 2.46 million kW. The completed investment for rural electricity network in the whole country was 6.5 billion Yuan; the newly increased capacity of 110 kV substation or above was 4.72 million kVA; the newly increased capacity of 35 (66) kV substation was 1.29 million kVA; the capacity of distribution transformer was 4.43 million kVA. The newly constructed 10 kV high pressure transmission line was 25,000 km and low pressure line was 43,000 km.

Soil and water conservation. In 2013, the allocated funds for soil and water conservation and ecological restoration projects under construction was up to 30.63 billion Yuan; the completed investment was 22.74 billion Yuan. The newly-added areas with soil conservation measures were up to 53,000 km^2, of which the areas under National Major Project for Soil Conservation were 13,200 km^2. The newly-added forest protected areas reached 16,800 km^2. There had been 2,153 small watersheds with comprehensive measures of soil and water conservation; 285 silt retention dams built on the Loess Plateau; and more than 2,000 landslides brought under control. The newly built basic farmlands were 727,000 ha (of which 553,000 ha were terrace lands and 16,000 ha were silted land). Newly-created forestland for soil conservation reached 1,411,000 ha and grassland 340,000 ha. More than 700 counties had been listed as national major project counties for soil and water conservation. The pilot project construction for erosion control in slope farmland extended to 160 counties in 22 provinces. The completed areas of slope for terraced field were 72,670 ha.

Capacity building. The completed investment for capacity building in 2013 was 6.11 billion Yuan, of which 1.95 billion Yuan was spent on procurement of communication equipment for flood control, 2.06 billion Yuan for hydrological facilities, 150 million Yuan for scientific research and education facilities, 1.67 billion Yuan for early-stage work, and 280 million Yuan for others.

III. Key Water Structures

Embankments and water gates. In 2013, the completed river embankments from Grade-I to Grade-V in the whole country had a total length of 276,800 km[1]. Of which, 179,800 km of embankment met the standard, with a percentage of 65.0% of the total up to standard. The length of embankment met the standard of Grade-I and Grade-II were 29,500 km, with a percentage of 76.8% of the total. These embankments can protect 570 million people and 43,000 ha of cultivated land. The number of water gates with a flow of 5 m^3/s increased to 98,192, of which 870 were large water gates, 7,985 flood diversion sluices, 17,509 drainage/return water sluices, 5,834 tidal barrages, 11,106 water diversion intakes and 55,758 controlling gates.

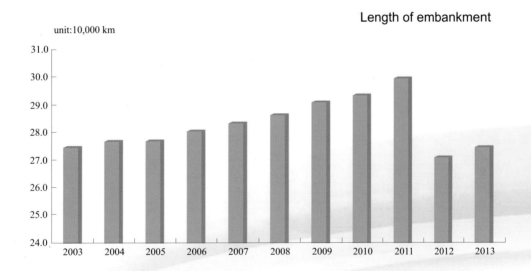

---

[1] The length of embankment before 2011 includes embankment below Grade-V.

Reservoirs and water complexes. The total number of reservoir in China boosted to 97,721, with a storage capacity of 829.8 billion m^3, of which 687 belong to large reservoirs with a total capacity of 652.8 billion m^3, accounting for 78.7% of the total; 3,774 medium-sized reservoirs with a total capacity of 107.0 billion m^3, accounting for 12.9% of the total. The percentage of large and medium reservoirs up to the safety standard ranked 96.8% of the total.

Irrigation. Irrigation districts with an area equal or above 2,000 mu added to 22,387, with a total cultivated land irrigated area of 33.928 million ha. In which, the irrigation districts equal or above 500,000 mu were 176, with a total cultivated land irrigated area of 6.241 million ha; the irrigation districts covering an area from 300,000 to 500,000 mu were 280, with a total cultivated land irrigated area of 5.010 million ha. By the end of 2013, the total cultivated land irrigated area reached to 63.473 million ha that accounted to 52.9% of the total cultivated area in China. The areas with water-saving irrigation facilities was totaled 27.109 million ha, among which 6.847 million ha equipped with sprinkler or micro irrigation systems and 7.424 million ha installed low-pressure pipes.

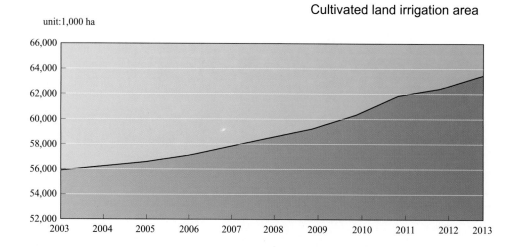

Cultivated land irrigation area

Tube wells and pumping stations. A total of 4.584 million water supply tube wells, with a daily water abstraction capacity equal or larger than 20 m^3 or an inner diameter larger than 200 mm, were excavated in the whole country. A total of 90,650 pumping stations that have an installed flow of 1 m^3/s or installed voltage above 50 kW had been installed, among which 366 belong to larger pumping stations, 4,161 medium-size and 86,123 small-size pumping stations.

Rural hydropower and electrification. By the end of 2013, hydropower stations built in rural areas totaled 46,849, with an installed capacity of 71.19 million kW, accounting for 25.4% of the total in China. The annual power generation by these hydropower stations reached to 223.3 billion kWh, accounting for 24% of the total power generation of the whole country, which helped 470,000 people being able to obtain electricity.

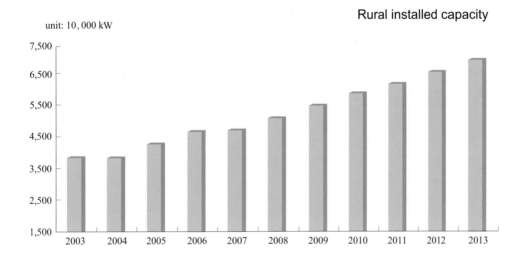

Soil and water conservation. In 2013, the restored eroded areas equaled to 1.0689 million km^2; ecological restoration areas accumulated to 770,000 km^2. A total of 160 ecologically-sound small watershed had been constructed. With the completion of check and acceptance of Phase-II of National Soil and Water Conservation Network and Information System, four center stations for river monitoring in the Huaihe, the Songliao, the Pearl and Taihu basins were built, including 18 master stations in 18 provinces, 75 monitoring substations and 715 monitoring points. Database systems for soil and water conservation had been built and developed.

Hydrology and informationization. By the end of 2013, the number of hydrological stations of all kinds increased to 86,554 in the whole country, including 3,195 national basic hydrologic stations, 816 special hydrologic stations, 9,330 gauging stations, 43,028 precipitation stations, 11,795 water quality stations, 16,407 groundwater monitoring stations, 14 evaporation stations and 57 experimental stations, and 1,912 soil moisture monitoring stations. China has built 24,518 telegram reporting stations and 1,207 hydrologic forecast stations. A total of 286 water environment monitoring centers (sub-centers) were put into operation that cover nearly all major rivers, lakes and reservoirs in China.

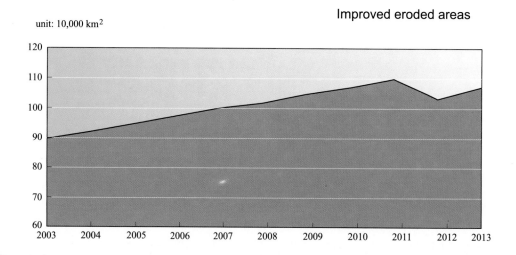

The water conservancy informatization which had opened a new era for an all round

and multi-layer development. The PCs and servers of varied kinds that connected to the internet reached 74,964 and 3,273 respectively. The on store capacity of equipment installed in the water resources departments at or above the provincial level reached 2,057,971.9 GB. A total of 105,930 information gathering points were installed for water departments at or above the provincial level, among which 63,461 were automatic information gathering points. Database under routine operation reached 706, with a storage capacity of 333,528.7 GB.

IV. Water Resources Utilization and Protection

According to preliminary statistics, the availability of water resources in 2013 totaled 2,795.79 billion m^3, 0.9% less than normal years. Mean annual precipitation was 661.9 mm, 3.0% more than normal years and 3.8% less than the year before. By the end of 2013, total water storage of 588 large reservoirs reached 300.54 billion m^3, 23.29 billion m^3 less compared with that of the beginning of the year; total water storage of 3,271 medium-size reservoirs were 39.53 billion m^3, 7.1 billion m^3 less than that of the beginning of the year.

In 2013, the total water supply amounted to 618.34 billion m^3, among which 81.0% came from surface water, 18.2% from underground aquifers and 0.8% from other water sources. The total water consumption increased to 618.34 billion m^3, of which domestic use amounted to 75.01 billion m^3 (in which urban domestic water use took 75.5%) or 12.1% of the total; industrial use 140.64 billion m^3 or 22.8% of the total; agricultural water use 392.15 billion m^3 or 63.4% of the total and environmental flow of 10.54 billion m^3 or 1.7% of the total. Comparing to that of the year before, the total water use increased by 5.22 billion m^3, of which domestic water use increased by 1.04 billion m^3, industrial use increased by 2.57 billion m^3, agricultural water use increased by 1.90 billion m^3 and environmental flow decreased by 0.29 billion m^3.

Water consumption per capita in 2013 was 456 m^3 in average. Water use of 10,000 Yuan GDP (at comparable price of the same year) was 109 m^3, a 7.6% decrease comparing with that in 2012. Water use of industrial production value added per 10,000 Yuan (at comparable price of the same year) was 67.0 m^3, 5.3% less comparing to that of the year before.

According to the result of water quality assessment on river sections of more than 208,000 km, rivers with better water quality that comply with or better than Class-Ⅲ standard occupied 68.6% of the total. Compared with that of 2012, the Class-Ⅰ, Class-Ⅱ, Class-Ⅲ is increased by 0.6%, the Class-V is decreased by 0.8%.

V. Flood Control and Drought Relief

Generally speaking, no large scale damages occurred as a result of flood and water-logging disasters in 2013. Nevertheless, a total of 11.901 million ha of cultivated land were affected by floods, resulting in 6,623 million ha damaged, 120 million people affected, 775 people dead, and 374 missing. A total of 530,000 houses were destroyed and 243 cities suffered from inundation. The disasters resulted in 314.6 billion Yuan of direct economic losses, among which the loss with water infrastructures reached 44.5 billion Yuan. The flood stricken areas include northeast parts of China and provinces of Sichuan and Guangdong etc. The death toll or the number of missing caused by mountain flood took 72% of the total in 2013. Economic loss caused by typhoon equaled to 40% of the total direct economic loss of flood and waterlogging disasters, which was the worst case since 2006.

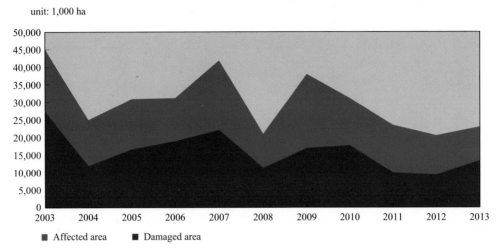

Flood or drought affected and damaged areas

In 2013, no large scale drought occurred in the whole country. However, some areas in the Yangtze River Basin, Huaihe River Basin, southern part of Yangtze River and Hanjiang River suffered from severe droughts. The farmland affected by droughts reached 11. 220 million ha, of which 6,971 million ha damaged, with a direct economic losses of 127. 5 billion Yuan. A total of 22. 41 million urban and rural population and 11. 79 million man-feed big animals and livestock suffered from temporary drinking water difficulties due to water shortage caused by these droughts.

In 2013, the funds allocated to flood control and drought relief amounted to 9. 889 billion Yuan, among which 592 million spent on structural measures, 2. 587 billion Yuan for extraordinary flood defense, 1. 626 billion Yuan for extraordinary drought relief, 4. 3 billion Yuan for mountain flood prevention and control and 784 million Yuan for emergency water diversion to ease the drought.

The flood control works protected 3. 967 million ha of cultivated land and 156 cities from flooding. The efforts of flood control and disaster reduction generated 235.8 billion Yuan of economic benefits. Safe drinking water was provided to 20.07 million people in rural and urban areas as well as 9.36 million big animals and livestock for alleviating temporary water shortage. The area with anti-drought measures reached 24.667 million ha that prevented a loss of 39.9 billion kg of grain.

VI. Water Management and Reform

Water resources planning and early-stage work. In 2013, there were 26 plans of varied kinds approved, among which 7 approved by the State Council, 6 approved by the National Development and Reform Commission (NDRC) and 13 approved by the Ministry of Water Resources (MWR). Preparation and revision of master plans of seven river basin commissions has made considerable progress. The revised master plans of seven major river basins were all approved by the State Council, including the master plans of Yellow River, Huaihe River, Haihe River, Peral River, Songhua River and Taihu Lake approved in 2013. Preparation work for compiling the National 13th Five-Year Plan on Water Development was initiated. National Plan for Modern Irrigation Development was entering appraisal stage. National long and medium term plans of balancing water supply and demand had achieved some results. Summary reports of National Plan for Water Resources Protection were finished.

The Implementation Plan for National Drought Relief (2014-2016) was submitted to the State Council for approval. Sand Excavation Plan of National Major River Courses was issued. The revision and approval of National Plan for Groundwater Utilization and Protection is still undertaking. Inspection and appraisals have been conducted for the National Plan of Soil and Water Conservation, National Plan for River Estuary and Seashore Development and Management and Water Distribution

Plan of Major River Basin (25 rivers in Phase - I). The National Plan for Water Development in Pasture Areas was completed. The formations of National Plan for Water Modernization Outlines and National Plan for Water-logging Control have been started.

A total of 46 projects were delivered to NDRC for approval, with a total investment of 146.149 billion Yuan. Among which NDRC approved 39, with a total investment of 107.733 billion Yuan, including 12 project proposals with a total investment of 42.806 billion Yuan, 18 feasibility studies with a total investment of 45.112 billion Yuan and 14 preliminary designs with a total investment of 28.831 billion Yuan.

Water legislation and administrative enforcement. In 2013, MWR approved/extended 1,420 administrative water permits, among which 119 qualification identifications for water resources assessment organizations; 2 water resources assessment reports of construction project; 355 qualification certificates of supervisors for water-related project construction; 159 qualification certificates of quality inspection of water-related project; 42 qualification certificates of supervisors on soil and water conservation of constructed project; 19 pre-approvals of environmental impact assessment reports of water and hydropower project; 285 approvals of soil and water conservation plan of production and construction projects; 179 check and acceptance of soil and water conservation plans of construction projects; 172 qualification certificates of survey and assessment for hydrology and water resources; and 72 licenses of headstock gear utilization.

In 2013, a total of 41 sand excavation zones were approved in the middle and lower reaches of the Yangtze River planned. There were 32 permits issued for allowing sand excavation business, with a total sand mining of 96.058 million ton. The sand mining boats that got licenses were 190. The investigated illegal cases totaled 55,437 and 50,036 or 90.26% of them resolved. The retrieved economic losses reached 182.99 million Yuan. A total of 5,537 water disputes resolved, and 48.15 million Yuan retrieved. There were 52 administrative reconsideration cases received by the Ministry of Water Resources that all being settled.

Water affairs management. So far, a total of 1,942 water affairs bureaus or water resources bureaus have been built at or above county level and assigned the responsibilities of water affairs management, which accounted for 79.7% of the total cities and counties. Among 1,479 established bureaus, 4 at provincial level, 7 at sub-provincial level, 213 at prefecture or city level, and 1,255 at county level. The utilities managed by water affairs bureaus own 4,039 water plants, 534,000 km of water supply pipes, with a daily water supply of 254.52 million m^3 and annual water supply of 37.96 billion m^3. A total of 2,159 sewage treatment plants were under operation, with a total pipeline of 263,000 km long and daily treatment capacity of 93.65 million m^3/d. The annual amount of sewage treatment reached to 25.41 billion m^3 in total.

A total of 2,761 water enterprises or companies under these water affair bureaus, with fixed assets equal to 238.96 billion Yuan, and annual income of 50.44 billion Yuan as well as a total profit of 430 million Yuan. The total investment putting into water industry in urban area amounted to 122.44 billion Yuan. Water sources for cities and counties totaled 4,566, with an annual water supply capacity of 112.40 billion m^3. The amount of recycled water use in cities (counties) increased to 4.82 billion m^3; while water supplied by unconventional sources despite of utilization of recycled water also added to 69.47 billion m^3.

Reform in project construction and management. The reform of national water project management system had completed and passed check and acceptance, which had reduced the number of water utilities within the scope of reform from 14,325 to 12,979 through merger or reorganization, with a decrease of 9%. A total of 12.96 billion Yuan had been used for covering the managerial staff and operation and maintenance cost of 12,979 water project management units, which covered 92% of the total cost. The expenses for operation and maintenance of public-good waterworks were 7.57 billion Yuan that covered 73% of the total cost. There were 8,995 organizations completed division of its type of operation, i.e. either totally self managed business or operated with government subsidy, accounted for 68.5% of the total water management units.

In 2013, another 22 enterprises got Class-A qualification of supervisors for water and hydropower project construction, 59 got the Class-B qualification and 187 got the Class-C qualification. Another 9 enterprises got Class-A qualification of supervisors for soil and water conservation project construction, 8 got the Class-B qualification and 20 got the Class-C qualification. No enterprise obtained Class-A qualification of supervisors for electromechanical and metal equipment manufacture and 2 obtained Class-B qualification. A total of 8 enterprises got the qualification (no grading is defined) of supervisors for environment protection of water project construction. In 2013, another 17 enterprises became Class-A quality inspection organizations for geotechnical engineering approved; 15 got the Class-A quality inspection organizations for concrete structures; 3 got the Class-A quality inspection organizations for metal structures; 1 Class-A quality inspection organizations for mechanical and electronic equipment; and 5 got Class-A quality inspection organizations for measuring and gauging tools. In 2013, the approved national water scenery spots reached 588, including 296 reservoirs, 117 natural rivers and lakes, 94 lake or riverine cities, 35 wetlands, 24 irrigation districts and 22 areas for soil and water conservation.

Reform in rural water resources management. The total number of Water User Associations (WUAs) established in the whole country reached to more than 80,500, and about 262 million mu of irrigated areas were under the operation of these associations that amounted to 28% of the total irrigated areas in China. A total of 29,000 grass-root water service organizations were crated in the whole county, with an employment of 130,000 people. Reform of management system had been continuously implemented in the irrigation districts, with the objectives of consolidating collection of "two fees", i.e. water resources fee and sewage treatment fee. The coverage rates of costs of basic personal expenses in public-good large irrigation districts or operation and maintenance of public waterworks reached to 60% and 41% respectively

Soil and water conservation. A total of 31,000 soil and water conservation plans of development and construction projects were examined and approved, of which 285 plans of national large and medium projects were approved by MWR, covering an area of 14,894 km^2. A total of 6,611 soil and water conservation projects completed check and acceptance. By the end of 2013, there were 12 provinces (autonomous regions or municipalities) promulgated the implementing provisions or regulations of Soil and Water Conservation Law. Soil and water conservation planning system has been popularized and extended to a larger scope.

Water pricing reform. In 2013, water supply for agricultural irrigation was at a cost of 0.2568 Yuan/m^3, among which the estimated cost of water supplied by state-owned enterprises was 0.1726 Yuan/m^3 and cost of water supply at the end-canal system was 0.0842 Yuan/m^3. Agricultural water price was 0.0914 Yuan/m^3, among which water price by state-owned enterprise was 0.0689 Yuan/m^3, water price at the end canal system was 0.0225 Yuan/m^3, which was about 35.59% of the actual cost. The collection rate of agricultural water charge was 86.48% in

average. The tariffs of water sources in these cities or counties ranged from 0.01 to 6.0 Yuan/m^3, among which water resources fees of surface water ranged from 0.002 to 2.0 Yuan/m^3 and water resources fees of groundwater resources ranged from 0.01 to 10.0 Yuan/m^3.

Reform of hydropower management system. Transformation of water use right on parable basis has been implemented in 17 provinces (autonomous regions or municipalities). Administrative rules and regulations for water-power resource development and utilization were promulgated in 16 provinces (autonomous regions or municipalities). It was stipulated by 15 provinces that water resources department shall take whole responsibility for integrating water resources management. Safety production has been reinforced for electric generation in rural areas by means of introducing approaches of "specifying person-in-charge of production and supervision", which had expanded the coverage of safety supervision to over 99%.

Production safety supervision. Generally speaking, safe production was realized in the water sector as no significant accident happened in 2013, except 15 production accidents and 24 people dead. There were 149 inspection teams that completed inspection of 429 construction projects. There were 2,229 problems identified and 301 notifications were released for correction of violating activities. Circulars were issued at national level to punish the severe violating activities of five projects in 5 provinces.

Re-inspection and supervision to 93 projects that have severe illegal activities were carried out by 31 inspection teams from river basin authorities. Moreover, a complete inspection and appraisal was conducted to 117 projects of 8 provinces which under the inspection of MWR since 2011. Guidance and leadership were provided to the inspection work of provincial water resources departments, who had completed inspection of 1,308 projects and found 9,215 illegal cases. A total of 960 rectify documents were issued.

Reservoir resettlement. According to uncompleted statistics, there were 114 large and medium reservoirs under construction in 2013, which involved 19 provinces (autonomous regions or municipalities), with a resettled population of 117,000, relocation sites of 367 and relocated rural residents of 94,000. The newly-constructed houses for resettlement covered an area of 3.3 million m^2 that need to occupy 96,000 mu of land.

Water science and technology. A total of 420 million Yuan had been allocated to science and technology projects, including 3 National Key Technology R&D Program being listed, 66 public-interest scientific research projects of the water sector, and 60 projects in "948 Plans", National Agricultural Science and Technology Achievements Transformation Fund Programs and MWR Key Technological Achievements Extension Plans. Four water technological project won the National Sci-Tech Advance Award. By the end of 2013, the numbers of national level or ministerial level labs were 12, and technical research centers were 13. Special funds for procurement and repairing of equipment of national scientific institutions amounted to 108.441 million Yuan. A total of 767 ministerial technical norms and standards are effective and 194 water-related technical standards are under review (revision). There were 81 technical norms and standards listed in the Table of Water-related Technical Standard System.

International cooperation. A total of 37 multilateral cooperation and exchange activities were successfully conducted, with 1 bilateral cooperation agreement signed. Under the fixed mechanism for bilateral exchange of government agencies, seven meeting were organized and held. The project of Key Plain and Low-lying Land Harness in Huaihe River Basin was under smooth progress, with a World Bank loan of 200 million US dollars. There were 5 governmental cooperation projects under implementation, 3 special projects for international cooperation and science and technology at national level got approval, having a funding of 8.97 million Yuan. There were 7 special projects for international cooperation and science and technology at national level under implementation, with a funding of 19.27 million Yuan.

VII. Current Status of Water Sector

Employees and salaries. In 2013, the employees of water sector were totaled 1.04 million, a 2.9% decrease comparing to that the year before. Of which the employees with long-term post amounted to 1.005 million, a 2.8% decrease. In the employees with long-term post, the staff working in the agencies directly under the Ministry of Water Resources was 70,000, a 5.4% decrease over the year before; the staff working in local agencies was 0.935 million, a 2.6% decrease. The total salary for the employees with long-term post in the whole country was 41.53 billion Yuan, a 6.7% increase comparing to that of 2012. The average salary per employee with long-term post was 41,453 Yuan, a 10.0% increase over 2012.

Employees and Salaries

	2003	2004	2005	2006	2007	2008	2009	2010	2011	2012	2013
number of in service staff/10^4 persons	122.9	118.2	110.5	109.2	106.8	105.6	103.7	106.6	102.5	103.4	100.5
of which, staff of MWR and agencies under MWR/10^4 persons	6.4	6.4	6.6	6.8	7.2	7.2	7.2	7.4	7.5	7.4	7.0
Local agencies /10^4 persons	116.5	111.8	103.9	102.3	99.6	98.4	96.5	96.3	95.0	96.0	93.5
salary of in-service staff/10^8 Yuan	140.6	157.1	159.8	184.3	211.28	234.37	264.74	297.91	351.37	389.1	415.3
average salary /(Yuan/person)	11,443	13,054	13,969	16,776	19,573	22,143	25,633	28,816	34,283	37,692	41,453

Reconnaissance and design. In 2013, the number of reconnaissance and design institutions obtained Class-A qualification increased to 108, and institutions awarded class-B qualification increased to 361 and institutions awarded Class-C qualification added up to 1,141, with a total staff of nearly 80,000 people.

Water Construction. The total number of water and hydropower construction companies awarded Super General Construction Contractor remained as 10. A total of 204 companies were approved to be Class-A Professional Contractors. A total of 38,767 people registered as supervisor engineers for water project construction and 13,525 people received Class- I Registered Certificate of Constructor in specialty of water and hydropower project.

Notes:

1. The data in this bulletin do not include those of Hong Kong, Macao and Taiwan.
2. The main index of national water resources development in 2012 has integrated with the data of First National Census for Water, but the data of the water-saving irrigated area in 2013 has integrated with the data of First National Census for Water.
3. The number of irrigation districts at 10,000 mu and its irrigated area is calculated based on the number of irrigation districts that have 10,000 mu of effective irrigated area or above in 2011. However, the statistics in 2012 was based on designed irrigated area that reached or upper to 10,000 mu.
4. Statistics of rural hydropower refers to the hydropower stations with an installed capacity of 50,000 kW or lower than 50,000 kW.

《2013年全国水利发展统计公报》编辑委员会

主　　　任：矫　勇
副　主　任：周学文
委　　　员：（按姓氏拼音排序）
　　　　　　蔡建元　陈大勇　陈茂山　段　虹　赫崇成　金　旸
　　　　　　匡尚富　李　戈　李原园　刘冬顺　倪文进　束庆鹏
　　　　　　汪安南　王杨群　王　治　许文海　杨得瑞　张新玉
　　　　　　赵　卫　祖雷鸣

《2013年全国水利发展统计公报》主要编辑人员

主　　　编：周学文
副　主　编：吴　强　黄　河
执 行 编 辑：杜国志　王　瑜　乔根平
主要参编人员：（按姓氏拼音排序）
　　　　　　杜国志　杜崇玲　巩劲标　荆茂涛　李　铭　吕　烨
　　　　　　倪　鹏　潘云生　齐兵强　乔根平　乔殿新　曲　鹏
　　　　　　田庆奇　王　瑜　吴钦山　徐　波　徐　岩　闫淑春
　　　　　　叶莉莉　张　范　张　岚　张淑娜　张晓兰
主要数据处理人员：（按姓氏拼音排序）
　　　　　　郭　悦　李　乐　齐　飞　王小娜

◎ 主编单位
水利部规划计划司

◎ 协编单位
水利部发展研究中心

◎ 参编单位
水利部办公厅
水利部政策法规司
水利部水资源司
水利部财务司
水利部人事司
水利部国际合作与科技司
水利部建设与管理司
水利部水土保持司
水利部农村水利司

水利部安全监督司
国家防汛抗旱总指挥部办公室
水利部农村水电及电气化发展局
水利部水文局（水利部水利信息中心）
水利部水库移民开发局
水利部南水北调规划设计管理局
水利部综合事业局
水利部水利水电规划设计总院
中国水利水电科学研究院